Twins for Optimizing Oil Extraction Processes

Copyright © 2024 DHIVAKAR POOSAPADI

All rights reserved. No part of this book may be reproduced or transmitted in any form or by any means, electronic or mechanical, including photocopying, recording, or by any information storage and retrieval system, without permission in writing from the copyright holder.

Table of Contents

Chapter 1: Introduction to Digital Twins in Oil Extraction 5

    Definition and Concept of Digital Twins 5

    Historical Context and Evolution 6

    Importance of Digital Twins in Oil Recovery 8

Chapter 2: Fundamentals of Oil Extraction Processes 11

    Overview of Oil Recovery Methods 11

    Challenges in Traditional Oil Extraction 13

    The Role of Technology in Modern Oil Extraction 16

Chapter 3: The Architecture of Digital Twins 18

    Components of Digital Twin Systems 18

    Data Acquisition and Integration 19

    Modeling Techniques and Simulation 21

Chapter 4: Application of Digital Twins in Oil Extraction 24

    Real-time Monitoring and Control 24

    Predictive Maintenance and Asset Management 26

    Optimizing Production Processes 27

Chapter 5: Case Studies in Oil Recovery Optimization 30

    Successful Implementations of Digital Twins 30

Lessons Learned from Industry Applications 32

Comparative Analysis of Traditional vs. Digital Twin Approaches 34

Chapter 6: Challenges and Limitations 36

Technical Limitations of Digital Twins 36

Data Management and Security Issues 38

Resistance to Adoption in the Industry 40

Chapter 7: Future Trends in Digital Twin Technology 43

Innovations in Digital Twin Development 43

Integration with Emerging Technologies 45

The Future of Oil Extraction and Sustainability 46

Chapter 8: Conclusion 49

Summary of Key Insights 49

Final Thoughts on Engineering Excellence 51

Call to Action for Industry Professionals 54

Chapter 1: Introduction to Digital Twins in Oil Extraction

Definition and Concept of Digital Twins

Digital twins represent a transformative technology that provides a virtual representation of physical assets, processes, or systems. In the context of oil extraction, a digital twin integrates real-time data and advanced simulation techniques to mirror the physical operations of oilfields, drilling rigs, and extraction processes. This dynamic digital counterpart allows stakeholders to analyze, predict, and optimize the performance of oil extraction activities by leveraging data analytics, machine learning, and artificial intelligence.

The concept of digital twins encompasses several key components, including data acquisition, integration, modeling, and visualization. Initially, data is collected from various sources, such as sensors installed on drilling equipment, geological surveys, and historical performance metrics. This data is then integrated into a cohesive model that accurately reflects the physical system. The modeling aspect is crucial, as it involves creating algorithms that can simulate the behavior of the oil extraction processes under varying conditions. Visualization tools further enhance this experience by providing intuitive representations of data and models, enabling engineers and decision-makers to interpret complex information effectively.

Digital twins are not static; they evolve alongside the physical assets they represent. This adaptability allows for real-time updates and adjustments based on the latest operational data.

For instance, if a drilling operation encounters unexpected geological formations, the digital twin can be recalibrated to assess the implications of this change on extraction efficiency and safety. This continuous feedback loop ensures that the virtual model remains relevant and useful for ongoing optimization efforts. The ability to simulate various scenarios and outcomes empowers engineers to make informed decisions that enhance productivity and minimize risks.

In the realm of oil recovery, digital twins facilitate the exploration of new strategies for resource extraction. By simulating different extraction techniques, engineers can evaluate their effectiveness without the need for costly and time-consuming physical trials. This capability is particularly valuable in complex reservoir conditions where traditional methods may not yield optimal results. The insights gained from digital twin simulations can lead to innovative approaches that improve recovery rates, reduce operational costs, and extend the lifespan of oilfields.

Ultimately, the definition and concept of digital twins extend beyond mere virtual models; they embody a paradigm shift in how engineers and professionals approach oil extraction processes. By leveraging the power of digital twins, the oil and gas industry can enhance operational excellence, drive efficiency, and adapt to the challenges posed by an ever-evolving energy landscape. As this technology continues to advance, its potential to revolutionize oil recovery practices will become increasingly apparent, establishing digital twins as an indispensable tool for industry professionals.

Historical Context and Evolution

The historical context of digital twins in the oil extraction industry can be traced back to the early days of computer modeling and simulations. In the 1960s and 1970s, advancements in computer technology allowed for the development of basic simulation models that represented physical processes in oil recovery. These early models predominantly focused on reservoir engineering and were primarily used to predict production rates based on geological data. The ability to use mathematical models to simulate reservoir behavior marked the beginning of a paradigm shift in the oil industry, paving the way for more sophisticated approaches to optimizing extraction processes.

As technology progressed, the 1980s and 1990s saw the introduction of more complex simulation software that incorporated various physical and chemical processes occurring during oil extraction. During this period, the need for real-time data became evident, leading to the integration of sensors and data acquisition systems into oil fields. These advancements enabled the collection of vast amounts of data, which could be analyzed to enhance the accuracy of simulation models. The development of these data-driven approaches laid the groundwork for the emergence of digital twins, which leverage real-time data to create dynamic, virtual representations of physical assets.

The evolution of digital twins in the oil and gas sector gained momentum in the early 2000s with the advent of the Internet of Things (IoT) and cloud computing. These technologies provided the necessary infrastructure to support the collection and processing of large datasets from various sources, including drilling rigs, production facilities, and environmental sensors. As a result, oil companies began to adopt digital twins to optimize their operations by simulating different extraction scenarios and predicting outcomes under varying conditions. This shift marked a significant transformation in how oil recovery processes were managed, moving from traditional methods to more agile, data-driven approaches.

In recent years, advancements in artificial intelligence and machine learning have further enhanced the capabilities of digital twins in oil extraction. These technologies allow for the analysis of complex datasets, enabling predictive maintenance, anomaly detection, and enhanced decision-making processes. Engineers and scientists can now create more accurate models that not only reflect current operational conditions but also adapt over time as new data is integrated. This evolution signifies a move toward a more holistic understanding of oil reservoirs, considering factors such as geological variations and equipment performance, thereby improving overall extraction efficiency.

The historical context and evolution of digital twins in oil extraction underscore the importance of interdisciplinary collaboration among engineers, scientists, and data analysts. As the oil and gas industry continues to face challenges related to sustainability and resource management, the integration of digital twins and simulation technologies becomes increasingly vital. By leveraging historical knowledge and embracing innovative technologies, professionals in the field can optimize oil recovery processes, ultimately contributing to more efficient and environmentally responsible extraction methods.

Importance of Digital Twins in Oil Recovery

The integration of digital twins in oil recovery processes has emerged as a transformative advancement within the industry. Digital twins serve as virtual replicas of physical assets, systems, or processes, enabling real-time monitoring, analysis, and optimization. In the context of oil recovery, these digital representations facilitate a deeper understanding of reservoir behavior, equipment performance, and production dynamics. By leveraging data from various sources, including sensors and historical performance metrics, engineers and scientists can create accurate simulations that inform decision-making and enhance operational efficiency.

One of the most significant advantages of utilizing digital twins in oil recovery is the ability to conduct predictive analytics. Engineers can model various scenarios and assess potential outcomes before implementing changes in the physical environment. This foresight minimizes risks associated with drilling operations and reservoir management, ultimately leading to more efficient resource allocation. By predicting how changes in pressure, temperature, and other variables affect oil recovery rates, organizations can optimize their strategies and improve overall recovery percentages.

Moreover, digital twins enhance collaboration among multidisciplinary teams involved in oil recovery. By providing a common platform for engineers, geologists, and data scientists, these tools foster a shared understanding of complex systems. Real-time data sharing allows for immediate feedback and iterative improvements, enabling teams to respond swiftly to challenges and opportunities. This collaborative approach not only accelerates innovation but also ensures that the best practices in engineering and science are consistently applied throughout the oil recovery process.

In addition to improving operational efficiency, digital twins contribute significantly to sustainability efforts in the oil and gas sector. By optimizing extraction processes and reducing waste, digital twins help organizations minimize their environmental footprint. Furthermore, they facilitate the transition to more sustainable practices by enabling the integration of renewable energy sources and carbon capture technologies. As the industry faces increasing regulatory pressures and societal expectations regarding environmental responsibility, the role of digital twins in promoting sustainable oil recovery becomes ever more critical.

Finally, the importance of digital twins in oil recovery extends beyond immediate operational benefits. As the industry evolves, the data generated and insights gained through digital twin technologies contribute to a broader understanding of subsurface environments and recovery techniques. This knowledge not only enhances current practices but also informs future research and development initiatives. By investing in digital twin technologies, organizations position themselves at the forefront of innovation in oil recovery, ensuring long-term competitiveness and resilience in a rapidly changing energy landscape.

Chapter 2: Fundamentals of Oil Extraction Processes

Overview of Oil Recovery Methods

The extraction of oil from subsurface reservoirs is a complex process that employs various recovery methods, each tailored to the geological and physical characteristics of the reservoir. These methods can be broadly categorized into primary, secondary, and tertiary recovery techniques. Primary recovery relies on natural reservoir pressure to bring crude oil to the surface, typically achieving a recovery factor of 5 to 15 percent of the oil in place. This method predominantly utilizes mechanical pumps and relies on the inherent energy of the reservoir, making it the least technologically demanding of the three categories.

Secondary recovery methods come into play once the natural pressure falls below viable levels for continued extraction. These techniques involve injecting water or gas into the reservoir to maintain pressure and enhance oil flow towards the production wells. Water flooding, the most common secondary method, involves injecting water into the reservoir to push oil towards extraction points. Gas injection, which may include natural gas or carbon dioxide, can also be employed to increase reservoir pressure and reduce oil viscosity. The implementation of these methods can significantly improve recovery factors, often reaching levels of 30 to 50 percent, thus extending the productive life of the reservoir.

Tertiary recovery, also known as enhanced oil recovery (EOR), is employed to extract remaining oil after primary and secondary methods have been exhausted. EOR techniques include thermal recovery, gas injection, and chemical flooding. Thermal methods, such as steam injection, reduce the viscosity of heavy oils, facilitating flow. Chemical flooding involves the injection of surfactants or polymers to improve the mobility of the oil, while gas injection methods, including CO_2 and nitrogen, are utilized to reduce oil density and increase pressure. Tertiary recovery can enhance recovery factors significantly, sometimes achieving total recoveries of 60 percent or more from the original oil in place.

The integration of digital twin technology into these recovery methods has begun to revolutionize the oil extraction process. Digital twins provide a virtual representation of physical systems, allowing engineers and scientists to simulate various recovery scenarios and optimize processes in real time. By analyzing data from the physical reservoir and correlating it with the virtual model, teams can predict outcomes of different recovery strategies, assess reservoir behavior under various conditions, and identify the most efficient methods for maximizing extraction. This dynamic modeling capability can lead to improved decision-making and significant cost savings over the lifecycle of oil recovery operations.

As the industry navigates the challenges of declining reserves and increasing environmental concerns, the role of digital twins in optimizing oil extraction processes becomes ever more critical. By leveraging advanced simulation techniques, engineers and researchers can develop innovative solutions that enhance recovery rates while minimizing ecological impacts. The future of oil recovery lies in the effective integration of these digital technologies, which promise to refine traditional methods and pave the way for more sustainable and efficient extraction practices.

Challenges in Traditional Oil Extraction

Traditional oil extraction methods face numerous challenges that can significantly impact efficiency and profitability. One of the primary concerns is the geological complexity of oil reservoirs. Variability in rock properties, fluid characteristics, and reservoir pressures complicates the prediction of oil behavior during extraction. Engineers often rely on historical data and geological models, yet the inherent uncertainties can lead to suboptimal decision-making. This unpredictability necessitates advanced modeling techniques, which could be enhanced through the application of digital twin technology, allowing for more accurate simulations of reservoir behavior under varying conditions.

Another critical challenge lies in the environmental and regulatory constraints surrounding traditional oil extraction. Stricter environmental regulations demand that companies not only minimize their ecological footprint but also adhere to comprehensive reporting requirements. The extraction process can lead to issues such as groundwater contamination and greenhouse gas emissions, which necessitate costly mitigation measures. Engineers must navigate these regulations while also striving to optimize extraction efficiency. Digital twins could facilitate compliance by enabling real-time monitoring and predictive analytics, allowing operators to address environmental concerns proactively.

Operational inefficiencies are also prevalent in traditional oil extraction methods. Aging infrastructure and equipment can lead to increased downtime, maintenance costs, and reduced output. The inability to predict equipment failures often results in unplanned outages, exacerbating the inefficiencies of the extraction process. By integrating digital twin technology, engineers can create virtual replicas of physical assets, allowing for predictive maintenance strategies that optimize operational performance and reduce costs. These digital models can provide insights into equipment health, enabling timely interventions before failures occur.

Furthermore, workforce challenges present significant hurdles for the oil extraction industry. There is a growing skills gap as experienced professionals retire and fewer new engineers enter the field. This shortage can lead to inefficiencies as companies struggle to maintain operational standards. Training programs often lag behind technological advancements, leaving workers ill-equipped to manage modern extraction technologies. Digital twins can play a pivotal role in addressing this issue by providing immersive training environments where engineers can interact with virtual models of extraction processes, enhancing their understanding and skills without the risks associated with real-world operations.

Lastly, the volatility of oil prices adds an additional layer of complexity to traditional extraction methods. Fluctuating market conditions can impact investment decisions and operational strategies, leading to uncertainty in project viability. This economic instability necessitates agile engineering solutions that can adapt to changing circumstances. Digital twins can help mitigate this challenge by allowing for scenario planning and sensitivity analysis, enabling engineers to model different market conditions and their impact on extraction strategies. With the ability to test various approaches virtually, organizations can make more informed decisions that align with market realities, ultimately improving their resilience in a fluctuating industry.

The Role of Technology in Modern Oil Extraction

The integration of technology in modern oil extraction has revolutionized the industry, enhancing efficiency and reducing costs. Advancements in digital technology, particularly the use of digital twins, have provided a platform for engineers and researchers to simulate and optimize extraction processes. These digital replicas of physical assets enable professionals to analyze real-time data, predict equipment behavior, and assess various operational scenarios without disrupting actual production. As a result, digital twins have become instrumental in refining extraction techniques and maximizing resource recovery.

One significant advantage of employing digital twins in oil extraction is their ability to facilitate predictive maintenance. By continuously monitoring the health and performance of extraction equipment, engineers can identify potential failures before they occur. This proactive approach minimizes downtime and maintenance costs, allowing for a more streamlined operation. The incorporation of artificial intelligence and machine learning algorithms further enhances the predictive capabilities of these systems, enabling a more sophisticated analysis of historical data and real-time inputs.

Moreover, digital twins enhance the decision-making process by providing a comprehensive understanding of reservoir behavior. Engineers can create detailed simulations that account for various geological and operational factors, allowing them to explore multiple extraction strategies without the risks associated with physical trials. This capability not only accelerates the development of effective recovery plans but also enables a more sustainable approach to resource management. By optimizing extraction methods based on simulation results, professionals can significantly reduce environmental impacts and improve the overall efficiency of operations.

Collaboration between interdisciplinary teams is another pivotal aspect of leveraging technology in oil extraction. Engineers, geologists, and data scientists can work together using digital twin technology to develop more integrated solutions. The shared platform fosters innovation and encourages the exchange of ideas, which can lead to breakthroughs in extraction techniques. Furthermore, the ability to visualize complex data in an intuitive manner enhances communication among team members, enabling more informed decision-making and fostering a culture of continuous improvement.

In conclusion, the role of technology, particularly digital twins, in modern oil extraction cannot be overstated. These tools not only enhance operational efficiency and predictive maintenance but also facilitate better decision-making and interdisciplinary collaboration. As the industry continues to evolve, the adoption of advanced technologies will be crucial in navigating the challenges of resource extraction while promoting sustainability. Engineers and researchers must remain at the forefront of these innovations, ensuring that the oil extraction processes are optimized for both economic and environmental considerations.

Chapter 3: The Architecture of Digital Twins

Components of Digital Twin Systems

Digital twin systems consist of several interconnected components that work cohesively to create a comprehensive virtual representation of physical assets. The primary element is the physical asset itself, such as an oil rig or a reservoir. This asset generates an immense volume of data during its operation, including operational parameters, environmental conditions, and performance metrics. The accurate capture of this data is crucial, as it serves as the foundation for the digital twin. Sensors and monitoring devices installed on the physical asset play a critical role in data acquisition, enabling real-time insights into the asset's status and performance.

The second component is the data management layer, which processes and stores the collected data. This layer ensures that the data is organized, accessible, and usable for analysis. Advanced data management techniques, including cloud storage and edge computing, facilitate efficient data handling and retrieval. Data quality and integrity are paramount, as they directly influence the reliability of the digital twin. Engineers and data scientists must implement robust data governance practices to ensure that the information feeding into the digital twin is accurate and up-to-date.

Next is the analytics and modeling component, which transforms raw data into actionable insights. Machine learning algorithms and simulation models are employed to analyze the data, identify patterns, and predict future behavior of the physical asset. This predictive capability is essential for optimizing oil extraction processes, as it enables engineers to anticipate equipment failures, optimize resource allocation, and enhance operational efficiency. The models can also simulate various scenarios, allowing for informed decision-making based on potential outcomes and risk assessments.

The visualization layer constitutes another critical component of digital twin systems. This layer presents the analyzed data in a user-friendly format, enabling engineers and decision-makers to interpret complex information easily. Visualization tools, such as dashboards and interactive 3D models, facilitate a deeper understanding of the asset's performance and health. Effective visualization enhances communication among stakeholders, allowing for collaborative problem-solving and strategic planning in oil recovery operations.

Finally, the feedback loop is integral to the functioning of digital twin systems. This component ensures continuous improvement by integrating insights gained from the digital twin back into the physical asset's operations. Through iterative learning, the digital twin can adapt to changes in operational conditions, thereby refining its predictions and enhancing overall system performance. This feedback mechanism not only helps in optimizing current processes but also informs future designs and operational strategies, positioning digital twin technology as a pivotal tool in the advancement of oil extraction methodologies.

Data Acquisition and Integration

Data acquisition and integration are pivotal components in the development and deployment of digital twins for optimizing oil extraction processes. The accuracy and effectiveness of digital twins hinge on the quality of data collected from various sources, including sensors, historical databases, and real-time monitoring systems. Engineers and scientists must ensure that the data acquisition process encompasses a wide range of parameters, such as reservoir characteristics, production rates, and environmental conditions. This comprehensive data collection facilitates a more nuanced understanding of the complex dynamics of oil recovery and allows for the creation of more accurate simulations.

The integration of diverse data sources presents both challenges and opportunities. Engineers must navigate the complexities associated with merging data from disparate systems, which often employ different formats, protocols, and standards. Establishing a unified data architecture is critical to streamline this process. Utilizing standardized data formats and APIs can significantly enhance interoperability, enabling seamless communication between various data sources. This integrated approach not only improves the reliability of the digital twin but also fosters collaboration among multidisciplinary teams, leading to better-informed decision-making.

Incorporating advanced technologies such as IoT devices and machine learning algorithms can further enhance data acquisition and integration efforts. IoT devices facilitate real-time data collection from the field, capturing vital information on pressure, temperature, and flow rates. Meanwhile, machine learning algorithms can analyze historical data to identify patterns and predict future performance. By leveraging these technologies, engineers can create more responsive digital twins that adapt to changing conditions in the field, allowing for proactive adjustments to extraction strategies.

Data validation and quality assurance are essential steps in the data acquisition process. Inaccurate or inconsistent data can lead to misguided analyses and suboptimal operational decisions. Engineers must implement robust validation protocols to ensure that the data collected is both accurate and representative of the actual conditions within the reservoir. Regular audits and cross-checks against established benchmarks can help maintain data integrity, thus enhancing the credibility of the digital twin and the simulations it generates.

Finally, the continuous feedback loop established through data acquisition and integration is crucial for the iterative improvement of digital twins. As new data becomes available from ongoing operations, it can be integrated into the existing models, allowing for refinements that reflect current realities. This dynamic approach not only optimizes the digital twin's performance but also fosters a culture of continuous learning and innovation within organizations. As engineers and researchers embrace these practices, the potential for enhanced oil recovery through digital twin technology becomes increasingly attainable, paving the way for more efficient and sustainable extraction processes.

Modeling Techniques and Simulation

Modeling techniques and simulation play a pivotal role in the development and optimization of digital twins in oil extraction processes. These methodologies enable engineers and researchers to create accurate representations of physical systems, allowing for the analysis of complex interactions in subsurface environments. By employing advanced modeling techniques such as finite element analysis, computational fluid dynamics, and machine learning algorithms, professionals can simulate various scenarios that affect oil recovery, ultimately leading to improved extraction strategies and resource management.

One of the primary modeling techniques used in digital twin development is the reservoir simulation model. These models typically incorporate geological, petrophysical, and fluid properties to simulate the behavior of hydrocarbons within the reservoir. By utilizing historical data and real-time monitoring, engineers can refine these models to predict future performance under varying operational conditions. This predictive capability is crucial for making informed decisions regarding drilling locations, production techniques, and enhanced oil recovery methods, ensuring that resources are utilized efficiently and sustainably.

In addition to traditional reservoir simulation, the integration of advanced technologies such as artificial intelligence and machine learning enhances the modeling process. These techniques facilitate the analysis of large datasets, enabling the identification of patterns and correlations that may not be immediately apparent through conventional methods. By leveraging AI-driven predictive analytics, professionals can optimize production schedules, reduce downtime, and enhance reservoir management strategies. The synergy between machine learning and digital twins leads to a more responsive and adaptive approach to oil extraction, ultimately improving overall operational efficiency.

The validation of modeling techniques through simulation is essential for ensuring accuracy and reliability. Engineers and scientists must continuously compare simulation results with empirical data to refine their models and enhance predictive capabilities. This iterative process not only strengthens the confidence in the digital twin's performance but also aids in identifying areas for improvement in extraction techniques. By engaging in rigorous validation, professionals can mitigate risks associated with oil recovery operations and make data-driven decisions that align with best practices in the industry.

As the oil extraction landscape evolves, embracing innovative modeling techniques and simulation frameworks will be critical for maintaining a competitive edge. The integration of digital twins into the oil recovery process represents a significant advancement in engineering practices, offering a comprehensive view of reservoir dynamics and operational performance. By fostering collaboration among engineers, scientists, and industry stakeholders, the ongoing refinement of these technologies will contribute to the sustainable and efficient extraction of oil, addressing the dual challenges of resource management and environmental stewardship.

Chapter 4: Application of Digital Twins in Oil Extraction

Real-time Monitoring and Control

Real-time monitoring and control are pivotal components in the optimization of oil extraction processes, particularly when integrated with digital twin technology. By leveraging real-time data, engineers can create dynamic models that accurately reflect the current state of oil extraction operations. These digital twins serve as a virtual counterpart to physical assets, enabling continuous updates based on live data streams from sensors and monitoring equipment. This capability allows for immediate response to operational changes, enhancing decision-making processes and ultimately leading to improved efficiency and reduced downtime.

The implementation of real-time monitoring systems involves the integration of advanced sensors and data acquisition technologies. These sensors collect critical information such as pressure, temperature, flow rates, and reservoir conditions. The data collected is transmitted to a centralized processing unit where it is analyzed and compared against the digital twin model. This real-time feedback loop allows engineers to identify deviations from expected performance, facilitating proactive adjustments to optimize extraction strategies. By maintaining a real-time connection between the physical and digital environments, organizations can ensure a more responsive and adaptable oil extraction operation.

In addition to operational efficiency, real-time monitoring and control significantly enhance safety and risk management within oil extraction processes. By continuously assessing the performance of equipment and the integrity of the extraction environment, potential hazards can be identified before they escalate into serious incidents. Digital twins equipped with predictive analytics can forecast equipment failures and suggest maintenance schedules based on actual wear and tear rather than arbitrary timelines. This predictive capability not only minimizes the risk of accidents but also reduces maintenance costs through more effective resource allocation.

Moreover, real-time monitoring systems facilitate improved collaboration among multidisciplinary teams involved in oil extraction. Engineers, geologists, and data scientists can access a shared digital twin platform that reflects the most current information about extraction processes. This shared visibility fosters enhanced communication and collaboration across teams, enabling them to make informed decisions collectively. The ability to visualize data in real-time empowers professionals to explore various scenarios and evaluate the impacts of different operational decisions, leading to better strategic outcomes.

Finally, the future of real-time monitoring and control in oil extraction processes is poised for further advancements with the integration of artificial intelligence and machine learning. These technologies can enhance predictive capabilities, enabling more sophisticated analyses of operational data. As digital twins evolve, they will incorporate advanced algorithms that not only monitor current conditions but also learn from historical data to improve future performance. This transformative approach will undoubtedly lead to more efficient, safe, and sustainable oil extraction practices, reinforcing the essential role of digital twins in the ongoing evolution of the industry.

Predictive Maintenance and Asset Management

Predictive maintenance and asset management are critical components of modern oil extraction processes, particularly when leveraging digital twin technology. The integration of predictive analytics into maintenance strategies allows for the anticipation of equipment failures before they occur, significantly reducing downtime and operational costs. This proactive approach relies on the continuous monitoring of equipment health through real-time data collection and analysis, enabling engineers to make informed decisions regarding maintenance schedules and resource allocation.

Digital twins play a pivotal role in this context by providing a virtual representation of physical assets, which facilitates detailed simulations of equipment performance under various operational scenarios. By analyzing data from sensors embedded in machinery, digital twins can predict potential failures based on historical performance patterns and current operating conditions. This predictive capability not only enhances the reliability of equipment but also optimizes the overall asset management process, ensuring that resources are utilized efficiently and effectively throughout the lifecycle of oil extraction operations.

The implementation of predictive maintenance models, supported by digital twin technology, allows companies to transition from traditional, reactive maintenance strategies to more sophisticated, condition-based approaches. This shift results in a more agile response to emerging issues, as maintenance activities can be scheduled precisely when needed rather than at predetermined intervals. Consequently, this leads to improved asset utilization and extended lifespan of critical equipment, which are particularly valuable in the highly competitive oil and gas sector.

Moreover, the synergy between predictive maintenance and asset management fosters a culture of continuous improvement within organizations. By harnessing the insights generated from digital twins, teams can refine their operational processes, identify inefficiencies, and develop targeted interventions that enhance productivity. This iterative feedback loop not only drives performance improvements but also supports strategic decision-making at higher organizational levels, aligning maintenance operations with broader business objectives.

Ultimately, the adoption of predictive maintenance and asset management practices, powered by digital twin technology, represents a significant advancement in the oil extraction industry. By leveraging data-driven insights, professionals can enhance the reliability, efficiency, and sustainability of their operations. As the industry continues to evolve, embracing these innovations will be essential for organizations looking to maintain a competitive edge in the face of increasing challenges and opportunities in the global energy landscape.

Optimizing Production Processes

Optimizing production processes in oil extraction is a critical endeavor that can significantly enhance efficiency, reduce costs, and minimize environmental impact. The integration of Digital Twin technology offers a transformative approach to understanding and refining these processes. By creating a virtual replica of the physical operation, engineers and scientists can simulate various scenarios, identify potential bottlenecks, and implement solutions in real time. This capability not only streamlines operations but also enables continuous monitoring and adjustment based on the latest data inputs, ensuring that production remains at peak efficiency.

The process of optimizing oil extraction begins with the collection and analysis of vast amounts of data from existing production systems. Digital Twins utilize this data to create a comprehensive model that mirrors the physical environment. By leveraging machine learning and advanced analytics, these models can predict outcomes of operational changes, allowing engineers to make informed decisions. For instance, adjustments in drilling techniques or extraction methods can be simulated to assess their impact on yield and resource utilization before any physical changes are made. This predictive capability is essential for maintaining a competitive edge in a rapidly evolving industry.

Furthermore, the application of Digital Twin technology extends beyond initial optimization. As oil fields age and conditions change, continuous monitoring through these digital models allows for timely adjustments to production strategies. By analyzing real-time data, engineers can detect anomalies, forecast equipment failures, and optimize maintenance schedules. This proactive approach not only enhances the lifespan of assets but also reduces unplanned downtimes, ultimately leading to more consistent production rates and higher overall profitability.

Collaboration among multidisciplinary teams is vital to fully exploit the potential of Digital Twins in optimizing production processes. Engineers, geoscientists, and data analysts must work in tandem to ensure that the models accurately represent the complex interactions within oil extraction systems. This collaborative approach fosters innovation, enabling teams to explore new methodologies and technologies that can further enhance efficiency. By sharing insights and expertise, organizations can develop more robust strategies for oil recovery, ensuring that they adapt to industry changes effectively.

In conclusion, optimizing production processes through the use of Digital Twins is an imperative for the future of oil extraction. As the industry faces increasing pressure to improve efficiency and sustainability, the ability to simulate, analyze, and adapt provides a crucial advantage. By harnessing the power of digital modeling and fostering collaboration among experts, the oil and gas sector can navigate the complexities of extraction while maximizing resource recovery and minimizing environmental impact. Embracing this technology will define the next generation of excellence in engineering and operational success.

Chapter 5: Case Studies in Oil Recovery Optimization

Successful Implementations of Digital Twins

The successful implementation of digital twins in the oil extraction industry exemplifies the transformative potential of this technology. Companies that have embraced digital twin models have reported significant improvements in operational efficiency, reduced costs, and enhanced decision-making capabilities. One notable case is the integration of a digital twin by a major oil and gas company in its offshore drilling operations. By creating a virtual replica of their drilling rigs, they were able to simulate various scenarios, optimize drilling parameters, and predict equipment failures before they occurred. This proactive approach not only minimized downtime but also maximized resource recovery, demonstrating the tangible benefits of digital twin technology in real-world applications.

In another instance, a leading operator successfully implemented digital twins to enhance reservoir management practices. The digital twin served as a comprehensive model that integrated geological data, production history, and real-time sensor inputs. This holistic view allowed engineers to conduct advanced simulations, leading to more accurate predictions of reservoir behavior under various extraction methods. As a result, the operator could make informed decisions on well placement and production strategies, significantly increasing recovery rates while minimizing environmental impact. This case highlights how digital twins can facilitate a deeper understanding of complex geological formations, ultimately driving more effective extraction processes.

Furthermore, the use of digital twins in predictive maintenance has revolutionized asset management in the oil sector. An exemplary implementation involved a mid-sized oil company that developed a digital twin for their pipeline infrastructure. By continuously monitoring the condition of pipes and utilizing machine learning algorithms, the digital twin could forecast potential failures and recommend maintenance schedules. This approach not only reduced the risk of environmental incidents but also optimized maintenance costs, ensuring that resources were allocated efficiently. The successful application of this technology illustrates the value of digital twins in maintaining the integrity and safety of critical infrastructure.

The deployment of digital twins in training scenarios has also yielded positive outcomes. A prominent oil and gas training institute incorporated digital twin technology into its curriculum, allowing trainees to interact with virtual models of complex drilling operations. This immersive experience provided students with a deeper understanding of the challenges faced in the field, enhancing their problem-solving skills and operational readiness. By bridging the gap between theoretical knowledge and practical application, digital twins have become an invaluable tool in the education and training of the next generation of engineers and professionals in the oil extraction domain.

Overall, the successful implementations of digital twins in various facets of oil extraction underscore their significance as a catalyst for innovation and efficiency. As the industry continues to evolve, the adoption of digital twin technology will likely expand, paving the way for further advancements in oil recovery methodologies. The lessons learned from these implementations serve as a roadmap for organizations looking to harness the capabilities of digital twins, ultimately leading to improved performance, sustainability, and economic viability in the oil extraction sector.

Lessons Learned from Industry Applications

The application of digital twins in oil extraction processes has revealed significant lessons that can enhance operational efficiency and decision-making. One of the primary lessons learned is the importance of real-time data integration. By utilizing a digital twin, operators can synchronize data from various sources, including sensors, historical records, and geological surveys. This integration allows for a comprehensive understanding of reservoir behavior, enabling engineers to make informed decisions based on real-time conditions rather than relying solely on historical data, which can often be outdated or misleading.

Another critical lesson is the value of predictive analytics within digital twin frameworks. By simulating different extraction scenarios, engineers can anticipate potential challenges and optimize production strategies. This proactive approach not only minimizes downtime but also maximizes resource utilization. The ability to run multiple simulations allows professionals to evaluate various operational parameters, such as pressure and temperature changes, ultimately leading to more effective management of extraction processes. This predictive capability helps organizations adapt to changing conditions in the field, reducing the risks associated with oil recovery.

Collaboration between multidisciplinary teams has also emerged as a vital lesson from industry applications of digital twins. The complexity of oil extraction requires input from geologists, reservoir engineers, data scientists, and IT specialists. By fostering a collaborative environment where these professionals can share insights and expertise, companies can leverage the full potential of digital twin technologies. Such collaboration ensures that the models created are not only technically sound but also aligned with business objectives, enhancing the overall effectiveness of oil extraction strategies.

Moreover, the experience gained from implementing digital twins highlights the necessity of continuous learning and adaptation. As technology evolves, so too must the methodologies used in oil extraction processes. Organizations have learned that maintaining flexibility in operations is crucial for incorporating new findings and advancements in digital twin technology. This adaptability allows companies to refine their approaches based on feedback from real-world applications, thereby improving the quality and accuracy of their digital models over time.

Finally, the importance of stakeholder engagement cannot be overstated. Successful digital twin implementations require buy-in from all levels of an organization, including management and field personnel. Engaging stakeholders early in the process not only facilitates smoother transitions to new technologies but also encourages a culture of innovation and accountability. Lessons learned from various industry applications demonstrate that when stakeholders are involved and informed, the adoption of digital twin technologies is more seamless, paving the way for enhanced efficiencies in oil extraction processes.

Comparative Analysis of Traditional vs. Digital Twin Approaches

The comparative analysis of traditional and digital twin approaches in the context of oil extraction processes reveals significant differences in methodology, efficiency, and overall impact on project outcomes. Traditional methods of oil recovery often rely on physical models, historical data, and empirical formulas to predict performance and optimize extraction. Engineers typically conduct extensive field tests and laboratory experiments to gather data, which can be time-consuming and costly. This approach, while foundational, often lacks the adaptability needed to respond to real-time changes in reservoir conditions, leading to potential inefficiencies and suboptimal decision-making.

In contrast, digital twin technology introduces a paradigm shift by creating a virtual replica of physical assets, processes, or systems. This approach allows for real-time monitoring and data integration from various sources such as sensors and IoT devices. Digital twins not only simulate the physical behavior of the oil extraction process but also provide predictive analytics that can forecast outcomes under different scenarios. This ability to analyze and visualize data in real time enables engineers to make informed decisions rapidly, reducing downtime and enhancing the overall efficiency of extraction operations.

Another crucial aspect of the comparative analysis is the scalability of both approaches. Traditional methods often require significant investment in physical infrastructure and personnel to conduct experiments and gather data on larger scales. In contrast, digital twins can be scaled up or down with relative ease, allowing for flexible simulations that accommodate changes in project scope or objectives. This scalability not only facilitates quicker iterations in the design and testing phases but also promotes a more agile response to market demands and operational challenges.

Moreover, the integration of machine learning and artificial intelligence within digital twin frameworks further enhances their capability to optimize oil extraction processes. Traditional approaches may struggle to process vast amounts of data effectively, while digital twins can leverage advanced algorithms to analyze patterns and generate insights that improve operational strategies. This analytical power enables predictive maintenance, optimized drilling techniques, and enhanced resource management, ultimately leading to increased recovery rates and reduced operational costs.

In conclusion, while traditional methodologies have laid the groundwork for oil extraction practices, the emergence of digital twin technologies offers a compelling alternative that addresses many of the limitations inherent in conventional approaches. The ability to simulate, predict, and adapt in real time positions digital twins as a transformative tool in the oil and gas industry. As engineers and scientists continue to explore the potential of digital twin applications, the insights gained from this comparative analysis will be instrumental in guiding future innovations and optimizing extraction processes.

Chapter 6: Challenges and Limitations

Technical Limitations of Digital Twins

Digital twins have emerged as a transformative technology in the field of oil extraction, enabling enhanced simulation, monitoring, and optimization of operations. However, several technical limitations hinder their full potential and effectiveness in this domain. One primary limitation lies in the fidelity of the models used to create digital twins. The accuracy of simulations depends heavily on the quality and granularity of the data fed into them. In many cases, the data may be incomplete or inconsistent due to variability in sensor readings, environmental conditions, or operational practices. This can lead to discrepancies between the digital twin and the physical asset, reducing the reliability of predictions and insights derived from the model.

Another significant challenge is the integration of disparate data sources. Oil extraction processes involve numerous subsystems, each generating vast amounts of data from various sensors and equipment. The lack of standardized protocols for data collection and management can create silos of information, making it difficult to create a cohesive digital twin that accurately reflects the entire operation. Additionally, the integration of legacy systems with modern digital twin technologies poses a technical hurdle. Engineers must navigate compatibility issues and potential data losses during the integration process, which can further complicate the development of a reliable digital twin.

Computational limitations also play a crucial role in constraining the effectiveness of digital twins in oil recovery applications. High-fidelity simulations often require significant computational resources, including advanced hardware and specialized software. The computational burden can lead to longer processing times, making real-time analysis and decision-making difficult. As a result, there is a trade-off between model complexity and the speed at which insights can be generated. Engineers must carefully balance the need for detailed simulations with the practical limitations of computing power, especially when quick responses are essential during critical operations.

Furthermore, the modeling of complex physical phenomena presents another technical limitation. Oil extraction involves intricate processes such as fluid dynamics, thermodynamics, and geomechanics, which can be challenging to model accurately. Simplifications made to these models for the sake of computational efficiency can lead to oversights or inaccuracies in the simulation outcomes. These inaccuracies may affect the optimization strategies derived from the digital twin, potentially leading to suboptimal decisions that could impact operational efficiency and safety. As such, ongoing research and development are necessary to enhance the modeling capabilities and address these complexities.

Finally, the human factor cannot be overlooked as a technical limitation in the deployment of digital twins. Engineers, scientists, and operators must possess the skills and knowledge to interpret the data produced by digital twins effectively. A lack of training or understanding of the underlying technologies can result in misinterpretation of the insights generated, leading to erroneous decisions. Consequently, fostering a culture of continuous learning and technical proficiency is vital for organizations looking to leverage digital twins in oil extraction processes. Addressing these technical limitations requires a concerted effort from the industry, emphasizing collaboration between engineers, researchers, and technology providers to drive innovation and improve the efficacy of digital twin applications.

Data Management and Security Issues

Data management and security are critical components of implementing digital twin technology in oil extraction processes. As the industry increasingly relies on data-driven decision-making, the volume and complexity of data generated from various sources, such as sensors, drilling equipment, and geological surveys, continue to grow. Effective data management practices are essential to ensure that this information is accurately captured, stored, and analyzed. Engineers and scientists must develop robust frameworks for data governance that encompass data quality, consistency, and accessibility, enabling seamless integration of information across different platforms and systems.

Moreover, the sensitive nature of the data involved in oil extraction processes necessitates stringent security measures. The risk of data breaches and cyberattacks is heightened in an environment where real-time data sharing and remote monitoring are prevalent. Professionals in the field must prioritize the implementation of advanced cybersecurity protocols to safeguard proprietary information and operational integrity. This includes employing encryption, secure access controls, and regular audits to detect vulnerabilities. By establishing a culture of security awareness and training personnel to recognize potential threats, organizations can significantly mitigate risks associated with data exposure.

In addition to cybersecurity concerns, the management of data throughout its lifecycle poses challenges that require attention. Digital twins rely on continuous data acquisition and real-time processing, making it imperative to maintain data integrity and consistency. Engineers and researchers must employ data validation techniques to ensure accuracy and reliability. This involves establishing protocols for data collection, storage, and processing that account for potential errors and discrepancies. Furthermore, the integration of artificial intelligence and machine learning algorithms can enhance data analytics capabilities, enabling more effective processing and interpretation of complex datasets.

Collaboration among stakeholders is vital in addressing data management and security issues in the context of digital twins for oil extraction. Engineers, scientists, and consultants should work together to develop industry standards and best practices that promote data sharing while ensuring compliance with regulatory requirements. Engaging in cross-disciplinary partnerships can facilitate knowledge exchange and foster innovative solutions to common challenges. By leveraging collective expertise, the industry can create a more resilient and secure data ecosystem that supports the sustainable development of oil extraction processes.

Finally, as digital twin technology continues to evolve, so too must the strategies for managing and securing data. Continuous research and development efforts are necessary to adapt to emerging threats and technological advancements. Professionals in the field should stay informed about the latest trends in data management and cybersecurity, participating in relevant training and professional development opportunities. By embracing a proactive approach to data management and security, engineers and researchers can enhance the effectiveness of digital twins in optimizing oil extraction processes, ultimately leading to improved operational performance and resource management.

Resistance to Adoption in the Industry

Resistance to the adoption of digital twin technology in the oil extraction industry has emerged as a significant barrier to realizing its full potential. This resistance is often rooted in a combination of cultural inertia, the complexity of existing processes, and the perceived risks associated with transitioning to new methodologies. Many professionals in the sector are accustomed to traditional practices that have developed over decades, leading to a reluctance to embrace innovative solutions that could disrupt established workflows. The challenge lies in overcoming this mindset to facilitate a smoother integration of digital twin technology.

One of the primary factors contributing to resistance is the lack of familiarity with digital twin concepts and their applications. Despite the proven benefits of digital twins in various sectors, including manufacturing and aerospace, the oil extraction industry remains hesitant. Engineers and scientists may find it daunting to learn new software tools and incorporate them into their existing processes, especially when they are already managing complex systems. This gap in knowledge can result in skepticism regarding the effectiveness of digital twins, making it imperative for proponents of the technology to invest in comprehensive training and education programs to demystify the technology.

Financial concerns also play a crucial role in the resistance to adoption. The initial investment required for implementing digital twin solutions can be perceived as a significant barrier, particularly in an industry that is often subject to fluctuating market conditions. Many organizations prioritize short-term returns on investment, leading to a reluctance to allocate resources toward technologies that may not yield immediate benefits. This short-sighted perspective can hinder long-term innovation and optimization, as the full potential of digital twins often becomes evident only after they have been integrated into operations over time.

Furthermore, the integration of digital twins necessitates a shift in organizational culture and collaboration across various departments. Traditional silos within organizations can impede the sharing of information and insights, which is essential for the successful implementation of digital twins. Engineers, scientists, and IT professionals must work collaboratively to ensure that data flows seamlessly between systems, enabling real-time monitoring and analysis. Resistance may arise from fears of altered job roles or the perceived threat of redundancy, which can create tension and hinder the collaborative spirit needed for successful adoption.

To address these challenges, industry leaders must champion the benefits of digital twin technology, showcasing successful case studies and demonstrating its capacity to enhance efficiency and reduce costs in oil extraction processes. By fostering an environment that values innovation and continuous improvement, organizations can begin to dismantle the barriers to adoption. Engaging stakeholders at all levels, from executives to field engineers, is essential in creating a shared vision for the future of oil extraction that embraces digital transformation through the integration of digital twins.

Chapter 7: Future Trends in Digital Twin Technology

Innovations in Digital Twin Development

The development of digital twin technology has seen significant innovations that enhance its application in optimizing oil extraction processes. These advancements are driven by the need for greater efficiency, accuracy, and sustainability in the oil industry. One notable innovation is the integration of real-time data analytics with digital twin models. By leveraging Internet of Things (IoT) sensors and advanced data acquisition systems, engineers can create dynamic models that reflect the current state of oil extraction operations. This allows for continuous monitoring and immediate adjustments based on real-time conditions, ensuring the optimization of recovery rates and resource management.

Another critical innovation is the incorporation of machine learning algorithms into digital twin frameworks. These algorithms enable predictive analytics that can forecast equipment failures, optimize maintenance schedules, and enhance decision-making processes. By analyzing historical data alongside real-time inputs, machine learning models can identify patterns and anomalies that may not be immediately apparent. This predictive capability not only minimizes downtime but also significantly reduces operational costs, thus contributing to more sustainable and efficient oil extraction practices.

Furthermore, advancements in simulation technologies have enhanced the fidelity of digital twins. High-fidelity simulations provide a more accurate representation of subsurface conditions and fluid dynamics, which are crucial for effective oil recovery. Engineers can simulate various extraction scenarios, assessing their potential impacts on production efficiency and environmental sustainability. This capability allows for better planning and risk management, as operators can visualize the effects of different strategies before implementation, leading to more informed decision-making.

Collaboration tools have also evolved, facilitating improved communication and information sharing among multidisciplinary teams. Cloud-based platforms enable engineers, scientists, and consultants to collaborate in real-time, regardless of geographic location. This interconnectedness enhances the development of digital twins by allowing diverse expertise to inform model creation and refinement. As professionals from various fields contribute to the digital twin lifecycle, the resulting models become more robust, addressing complex challenges in oil extraction with greater efficacy.

Lastly, the focus on sustainability has spurred innovations that integrate environmental considerations into digital twin development. Engineers are now designing models that not only optimize extraction processes but also assess the ecological impact of operations. By simulating various extraction techniques, digital twins can identify environmentally friendly practices that minimize carbon footprints and enhance regulatory compliance. This dual focus on operational efficiency and sustainability positions digital twins as a pivotal tool in the future of oil extraction, aligning industry practices with the global push for sustainable development.

Integration with Emerging Technologies

The integration of emerging technologies with digital twins is pivotal for optimizing oil extraction processes. Digital twins, which serve as virtual replicas of physical systems, enable real-time data analysis and simulation, allowing engineers to make informed decisions. As the oil and gas industry faces increasing pressure to enhance efficiency and sustainability, the incorporation of advanced technologies such as artificial intelligence, machine learning, and the Internet of Things is becoming essential. These technologies not only improve the accuracy of digital twins but also facilitate predictive maintenance, operational efficiency, and resource management.

Artificial intelligence and machine learning algorithms can analyze vast datasets generated by equipment and sensors used in oil extraction. By processing this information, engineers can identify patterns and anomalies that may indicate potential issues before they escalate. This predictive capability not only minimizes downtime but also enhances the overall reliability of extraction operations. Moreover, machine learning models can continuously improve their predictions as more data becomes available, leading to increasingly optimized processes over time.

The Internet of Things (IoT) plays a crucial role in enhancing the functionality of digital twins by providing real-time data from various sources, such as drilling equipment, pipelines, and environmental sensors. This interconnectivity allows for a comprehensive view of operations, enabling engineers to monitor performance metrics and environmental conditions closely. By integrating IoT devices with digital twin technology, professionals can simulate different scenarios and assess the potential impacts of various operational strategies, leading to more informed decision-making and risk management.

Blockchain technology also presents valuable opportunities for enhancing digital twin applications in oil extraction. By ensuring secure and transparent data sharing among stakeholders, blockchain can facilitate better collaboration and trust in the data generated by digital twins. This is particularly important in an industry where multiple parties are often involved in exploration and production activities. The integration of blockchain can streamline processes such as supply chain management and regulatory compliance, ultimately contributing to more efficient and accountable operations.

In conclusion, the integration of emerging technologies with digital twins is transforming the landscape of oil extraction processes. By leveraging AI, machine learning, IoT, and blockchain, engineers and professionals can optimize operations, enhance predictive capabilities, and foster collaboration among stakeholders. As the industry continues to evolve, the effective use of these technologies will be crucial for achieving engineering excellence and ensuring sustainable practices in oil recovery. The future of oil extraction lies in the ability to harness these innovations to create smarter, more efficient, and resilient systems.

The Future of Oil Extraction and Sustainability

The future of oil extraction is increasingly intertwined with the principles of sustainability, driven by both technological advancements and environmental imperatives. As global energy demands continue to rise, the oil and gas industry faces the dual challenge of maximizing resource recovery while minimizing ecological impacts. Digital twin technology stands at the forefront of this evolution, enabling engineers and researchers to create precise virtual models of extraction processes. These models facilitate real-time monitoring and optimization, allowing for more efficient use of resources and a significant reduction in waste and emissions.

Digital twins provide a comprehensive framework for simulating various extraction scenarios, analyzing the potential outcomes of different operational strategies. By integrating data from sensors and historical performance, these models can predict how changes in extraction techniques or equipment will affect efficiency and sustainability. This predictive capability empowers professionals to make informed decisions that enhance oil recovery while adhering to stringent environmental standards. The continuous feedback loop established by digital twins ensures that operational adjustments can be made swiftly, responding to both market demands and regulatory requirements.

Moreover, the adoption of advanced analytics and artificial intelligence within the digital twin framework further augments its effectiveness in promoting sustainable practices. Machine learning algorithms can identify patterns and anomalies in extraction processes, leading to proactive maintenance and the prevention of costly downtimes. This not only optimizes production but also minimizes the environmental footprint associated with unexpected failures. As such, professionals in the field must embrace these technologies, as they represent a significant leap towards more sustainable oil extraction methodologies.

Collaboration among engineers, scientists, and industry stakeholders will be essential in shaping the future landscape of oil extraction. By fostering a culture of innovation and knowledge sharing, the industry can leverage cross-disciplinary insights to develop more sustainable practices. Initiatives that promote the integration of digital twins with renewable energy sources and carbon capture technologies will be vital in transitioning toward a more sustainable energy paradigm. This collaborative approach will not only enhance operational efficiency but also build resilience against the fluctuating dynamics of global energy markets.

In conclusion, the future of oil extraction will be defined by the successful integration of digital twin technology and sustainable practices. As the industry adapts to the growing demands for environmental responsibility, engineers and researchers must remain at the forefront of this transformation. By harnessing the full potential of digital twins, the oil and gas sector can optimize extraction processes, reduce environmental impacts, and pave the way for a more sustainable energy future. Embracing these advancements will ensure that oil extraction remains viable and responsible in an increasingly eco-conscious world.

Chapter 8: Conclusion

Summary of Key Insights

The integration of digital twins into oil extraction processes represents a significant advancement in the optimization of resource recovery. By creating virtual replicas of physical assets, engineers can simulate the behavior of oil fields under various conditions, allowing for more informed decision-making. This approach not only enhances understanding of complex systems but also enables the anticipation of potential issues before they arise. The ability to conduct real-time monitoring and predictive analysis through digital twins allows for improved operational efficiency and reduces downtime, ultimately leading to cost savings and increased production rates.

One of the key insights from the exploration of digital twins in oil extraction is the importance of data integration and analytics. The successful deployment of digital twin technology relies heavily on the availability and quality of data from various sources, including geological, environmental, and operational parameters. Engineers and scientists must focus on developing robust data acquisition systems and advanced analytical tools to ensure that the digital twin accurately reflects the real-world processes. The synergy between data science and engineering practices is critical in deriving actionable insights that can inform strategic planning and resource management in oil recovery operations.

Another significant takeaway is the role of collaboration and interdisciplinary approaches in maximizing the benefits of digital twin technology. The complexity of oil extraction processes necessitates a team of professionals from diverse fields, including geoscience, engineering, computer science, and data analytics. By fostering a collaborative environment, teams can leverage their collective expertise to develop more sophisticated models and simulations. This holistic approach not only enhances the accuracy of digital twins but also encourages innovative solutions to longstanding challenges in oil recovery, such as reservoir management and enhanced oil recovery techniques.

Additionally, the implementation of digital twins can lead to more sustainable practices in the oil industry. As the sector faces increasing pressure to reduce its environmental footprint, digital twins can facilitate more efficient resource use and minimize waste. By simulating various extraction scenarios, engineers can identify the most effective strategies for maximizing yield while adhering to environmental regulations. This proactive stance not only helps in meeting compliance requirements but also positions companies as leaders in sustainable practices, appealing to stakeholders and investors who prioritize environmental responsibility.

Lastly, the ongoing evolution of digital twin technology and its applications in oil extraction highlights the necessity for continuous learning and adaptation within the industry. As technological advancements emerge, professionals must stay abreast of the latest developments in simulation, machine learning, and artificial intelligence. Investing in training and professional development will be essential for engineers and scientists to effectively harness these innovations and drive the future of oil recovery. The insights gained from this exploration of digital twins underscore the transformative potential of technology in optimizing oil extraction processes and the critical role of skilled professionals in realizing these advancements.

Final Thoughts on Engineering Excellence

In the rapidly evolving landscape of oil extraction, the implementation of digital twins represents a transformative approach to enhancing operational efficiency and optimizing resource management. Engineers and scientists are increasingly recognizing that the integration of this technology can lead to significant advancements in the accuracy of simulations, ultimately driving better decision-making processes. Digital twins offer a real-time, data-driven reflection of physical assets, enabling professionals to monitor and analyze the performance of oil extraction operations with unprecedented precision. As we conclude our exploration of engineering excellence in this domain, it is crucial to appreciate the profound implications that digital twins hold for the future of oil recovery.

The convergence of digital twin technology with traditional engineering practices encourages a culture of continuous improvement and innovation. By leveraging advanced analytics and machine learning, engineers can uncover insights that were previously inaccessible, allowing for proactive adjustments in extraction strategies. This empowers teams to mitigate risks and enhance the sustainability of operations, aligning with industry goals for environmental stewardship. The ongoing refinement of digital twin models not only supports operational excellence but also fosters a more resilient framework capable of adapting to the inherent uncertainties of the oil market.

Collaboration emerges as a pivotal theme in achieving engineering excellence through digital twins. The interdisciplinary nature of oil recovery necessitates that engineers, scientists, and consultants work together, sharing expertise and insights to refine simulation models. This collaborative approach can lead to the development of robust frameworks that incorporate diverse perspectives, enhancing the reliability of predictions and strategies. As professionals engage in knowledge-sharing, they contribute to a collective intelligence that drives the industry forward, paving the way for innovative solutions to complex challenges.

Moreover, the continuous evolution of technology in this field cannot be overlooked. As computational power increases and data acquisition techniques advance, the capabilities of digital twins will expand, offering even more sophisticated models for oil recovery. Engineers must remain vigilant and adaptable, keeping abreast of technological advancements that could influence their work. Embracing a mindset of lifelong learning will be essential for professionals seeking to maintain their relevance and effectiveness in a landscape that is constantly shifting.

Ultimately, the pursuit of engineering excellence through the application of digital twins in oil extraction is not merely an aspiration but a necessity for the industry's future. By harnessing the potential of this innovative technology, professionals can significantly improve operational efficiency, reduce costs, and enhance sustainability. As we reflect on the insights shared throughout this discourse, let us commit ourselves to fostering a culture of excellence, collaboration, and innovation, ensuring that the oil recovery sector thrives in an increasingly complex and competitive environment.

Call to Action for Industry Professionals

As the oil extraction industry faces increasing pressure to enhance efficiency and reduce environmental impact, the integration of Digital Twin technology offers a transformative opportunity for professionals in the field. Engineers, scientists, and researchers must recognize the potential of these dynamic digital representations of physical assets, processes, and systems. By embracing Digital Twins, industry professionals can optimize oil recovery processes, improve decision-making, and drive innovation within their organizations. The time has come for industry leaders to take decisive action and champion the adoption of this technology.

To effectively implement Digital Twin solutions, professionals should prioritize collaboration across disciplines. The complexity of oil extraction processes requires a multidisciplinary approach, combining expertise from engineering, data science, and environmental studies. By fostering partnerships between these fields, engineers and researchers can develop comprehensive models that accurately simulate real-world scenarios. This collaborative spirit will not only enhance the understanding of oil recovery systems but also facilitate the identification of best practices and innovative strategies for optimization.

Furthermore, professionals in the oil extraction sector must invest in continuous education and training to stay abreast of advancements in Digital Twin technology. As the field evolves, so too do the tools and methodologies associated with it. Industry experts should seek out training opportunities, workshops, and conferences that focus on the latest developments in simulation and modeling. By enhancing their skills and knowledge base, engineers and scientists can position themselves as leaders in the adoption of Digital Twins, ensuring that their organizations remain competitive in a rapidly changing landscape.

Another critical aspect of this call to action is the need for a cultural shift within organizations. Embracing Digital Twin technology requires a willingness to innovate and adapt to new ways of working. Industry professionals must advocate for a mindset that values data-driven decision-making and iterative processes. This cultural transformation will empower teams to experiment with simulation and modeling, ultimately leading to more effective oil recovery strategies. Leadership plays a pivotal role in this shift, and it is essential for professionals to engage with management to promote the benefits of Digital Twins and secure the necessary resources for implementation.

Finally, industry professionals should actively participate in the broader conversation surrounding Digital Twin applications in oil extraction. Engaging with peers through forums, publications, and collaborative projects will not only share knowledge but also contribute to the collective advancement of the field. By showcasing successful case studies and sharing insights, professionals can inspire others to explore the potential of Digital Twins. The future of oil recovery depends on the commitment of engineers, scientists, and consultants to champion this innovative technology and leverage it to achieve engineering excellence in their practices.